MATH GENIUS

Journal and Reference Guide

by the Summer 2020 Math Team
Rania Shalaby, Kim Hunter, Cherie Brown, Leya Booth, and Kim Evans
Innovative Education Management

Genius Academic Press
Los Angeles, California

Innovative
Education
Management

Math Genius Homework Journal and Reference Guide

Published By:
Genius Academic Press, an imprint of Genius Books & Media, Inc.
31858 Castaic Road, #154
Castaic, CA 91384

Edited By: Rania Shalaby, Kim Hunter, Cherie Brown, Leya Booth, and Kim Evans

Special Thanks to:
Innovative Education Management
South Sutter
Sky Mountain
Ocean Grove

ISBN: 978-1-947521-39-1

200729

Math Journal

Key Term:

Definition:

Examples and Illustrations:

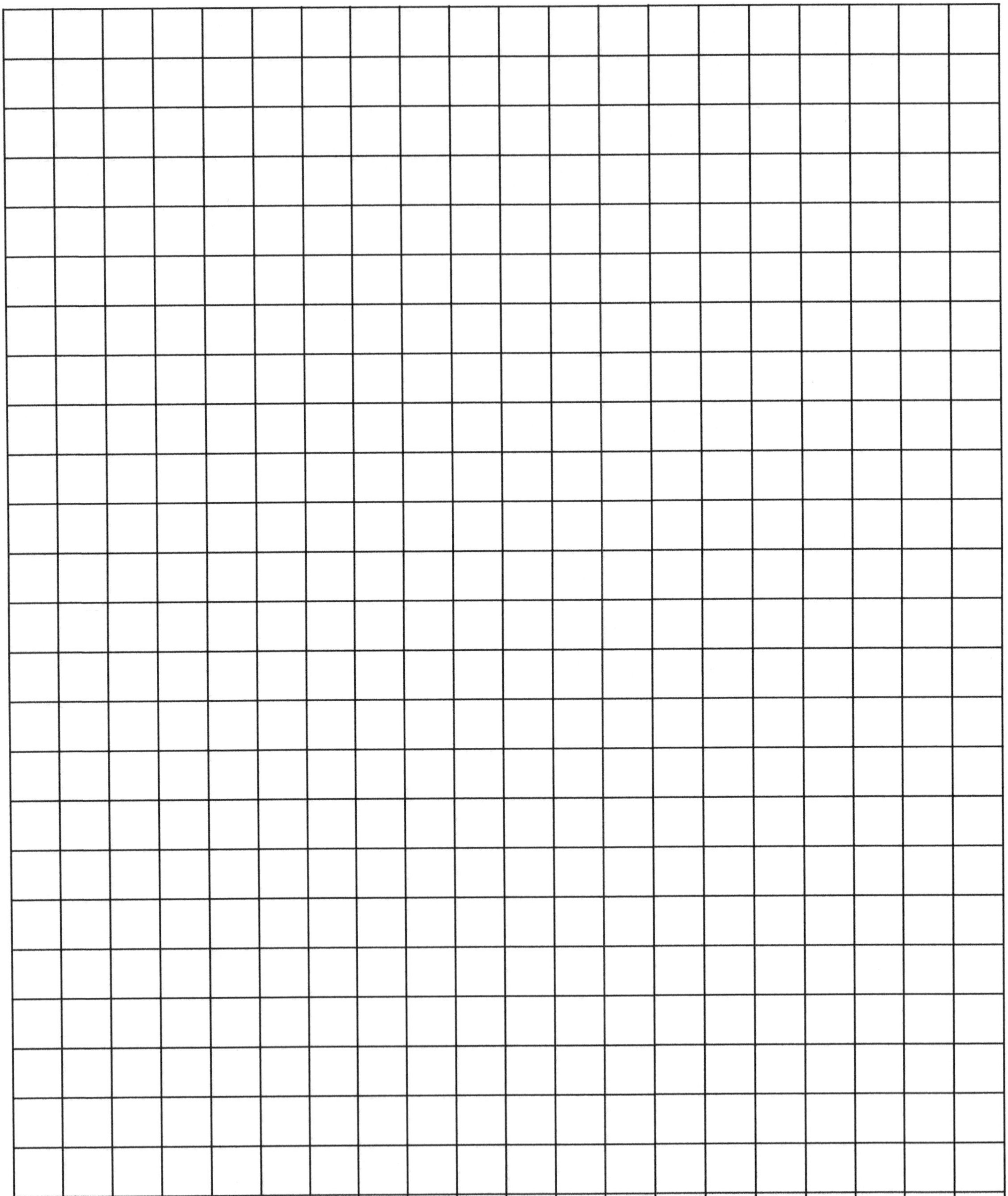

Math Journal

Key Term:

Definition:

Examples and Illustrations:

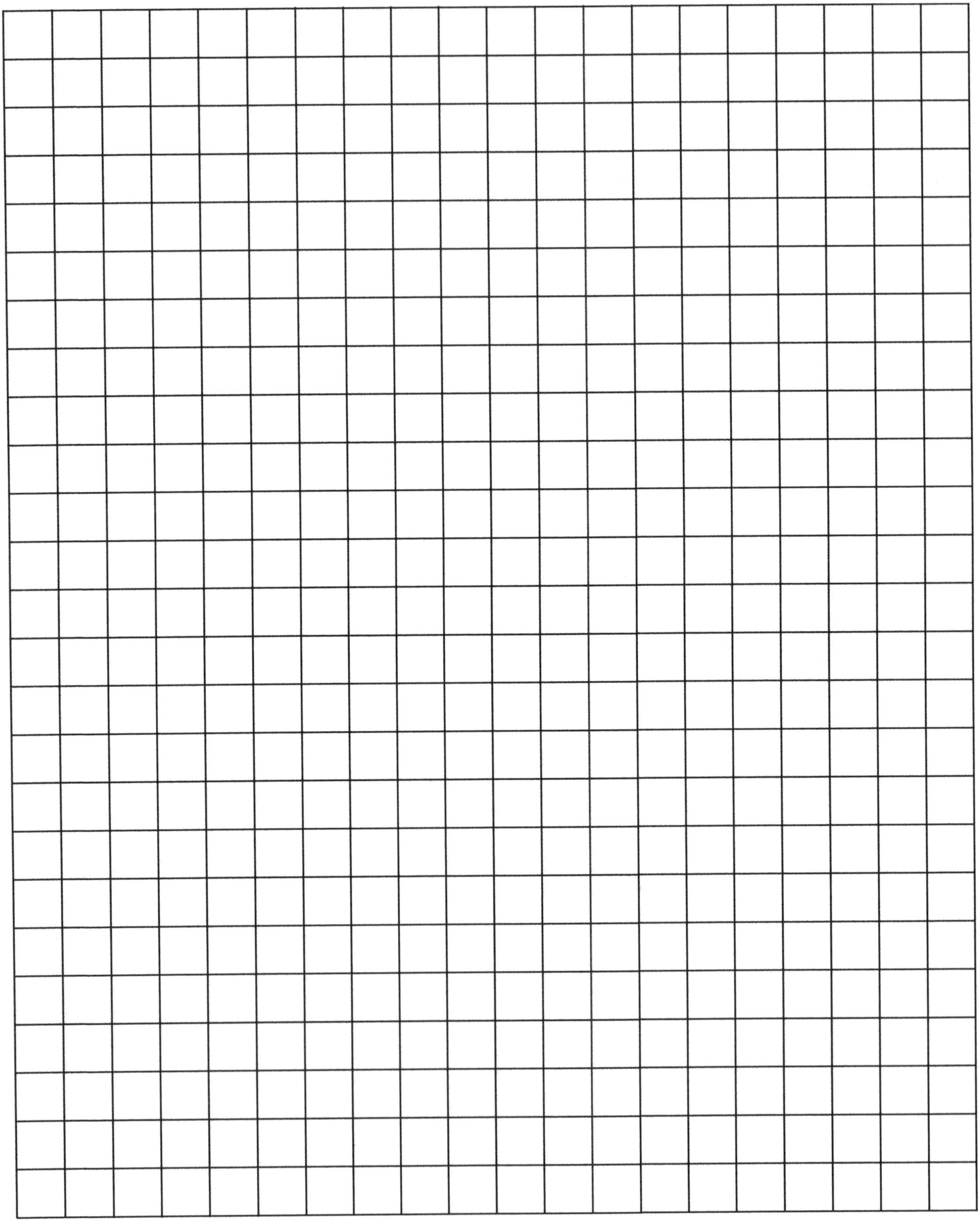

Math Journal

Key Term:

Definition:

Examples and Illustrations:

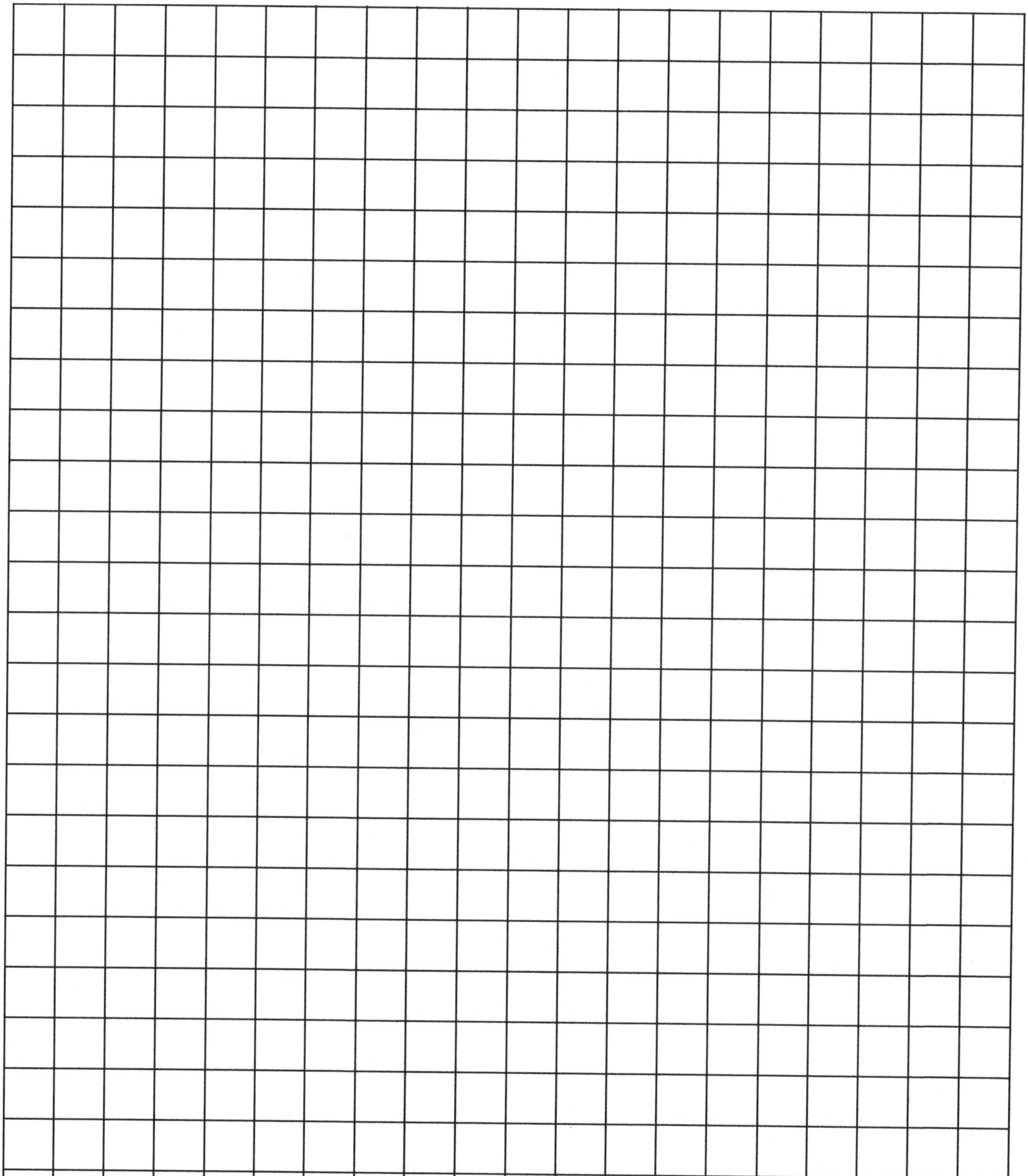

Math Journal

Key Term:

Definition:

Examples and Illustrations:

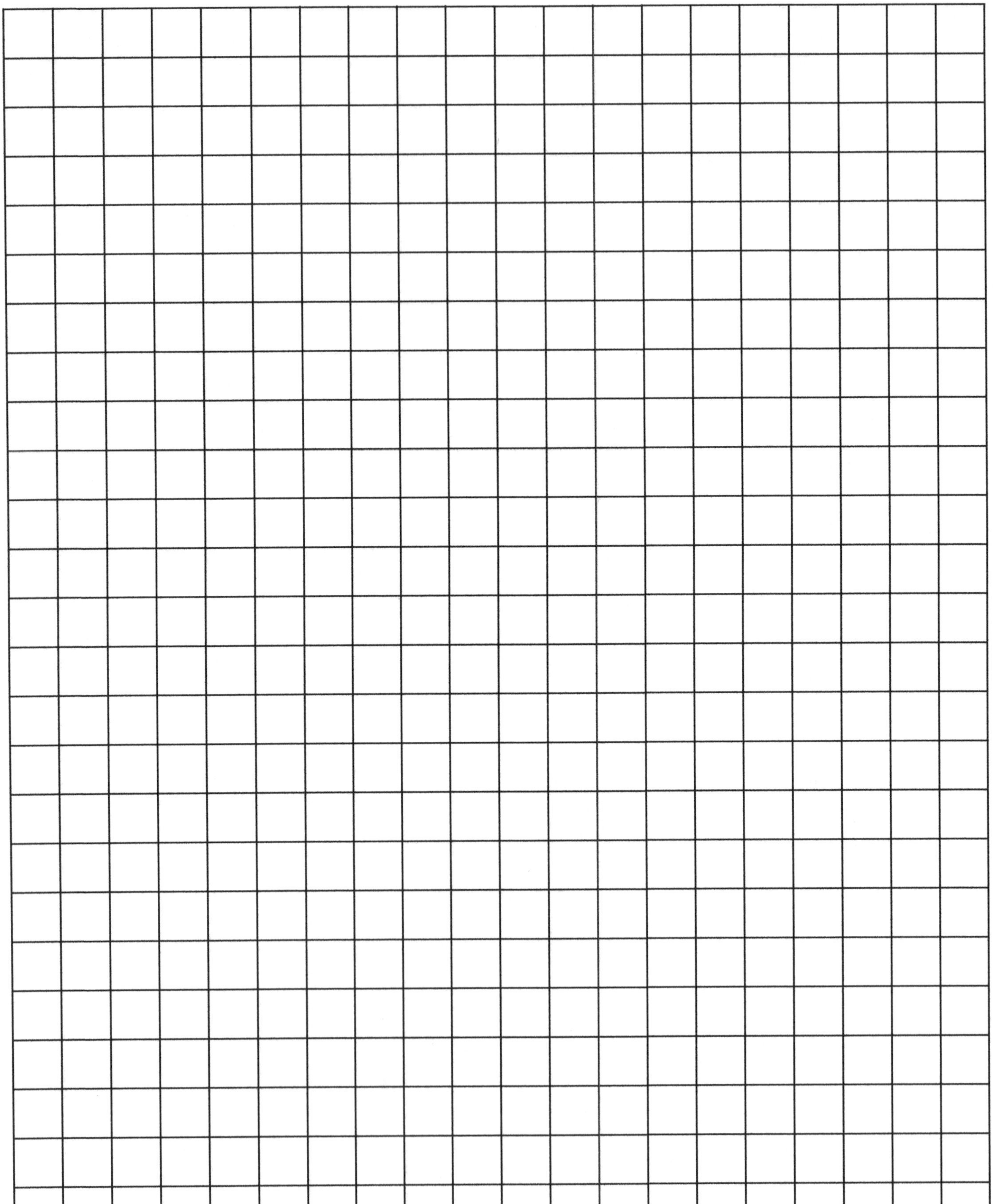

Math Journal

Key Term:

Definition:

Examples and Illustrations:

Math Journal

Key Term:

Definition:

Examples and Illustrations:

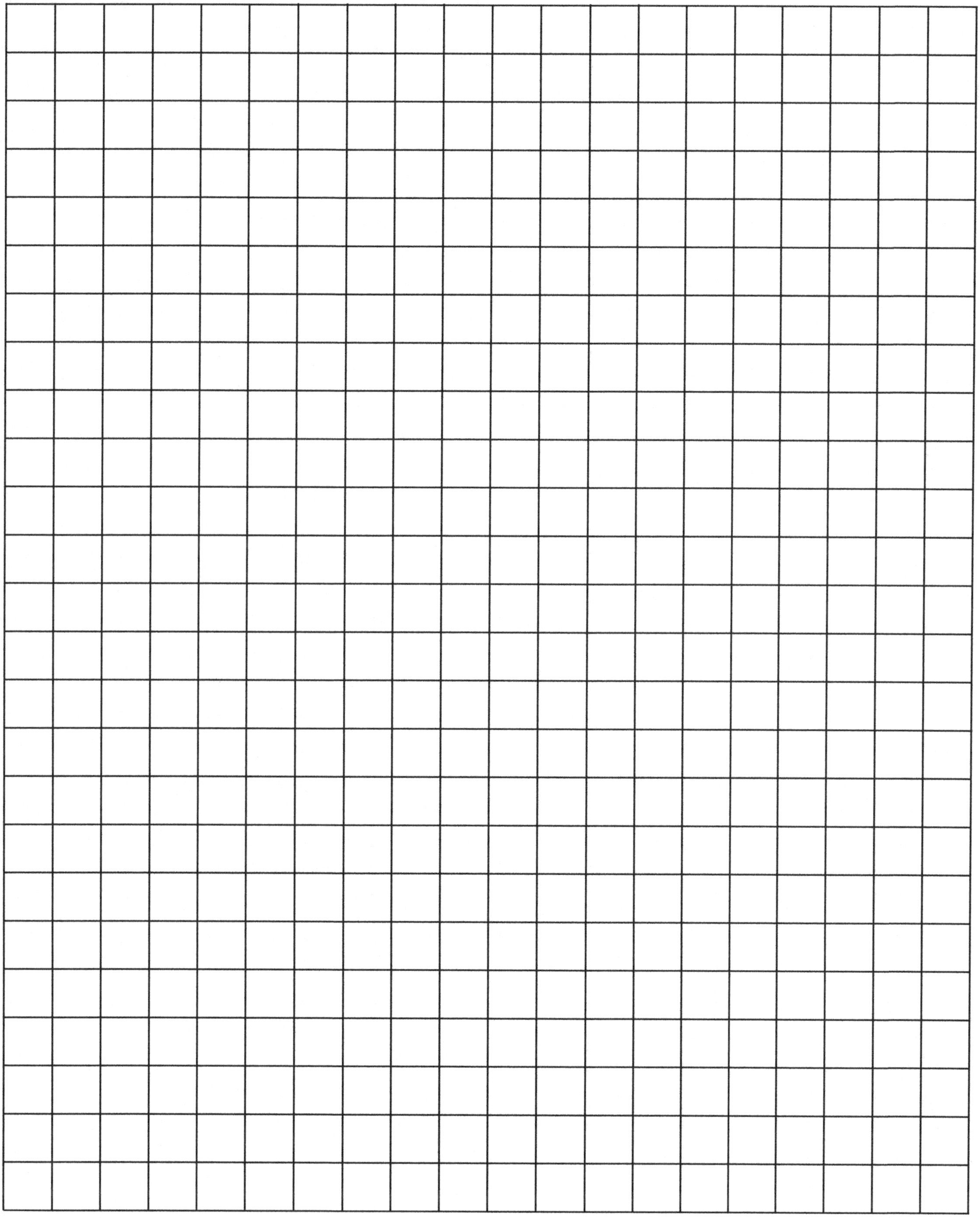

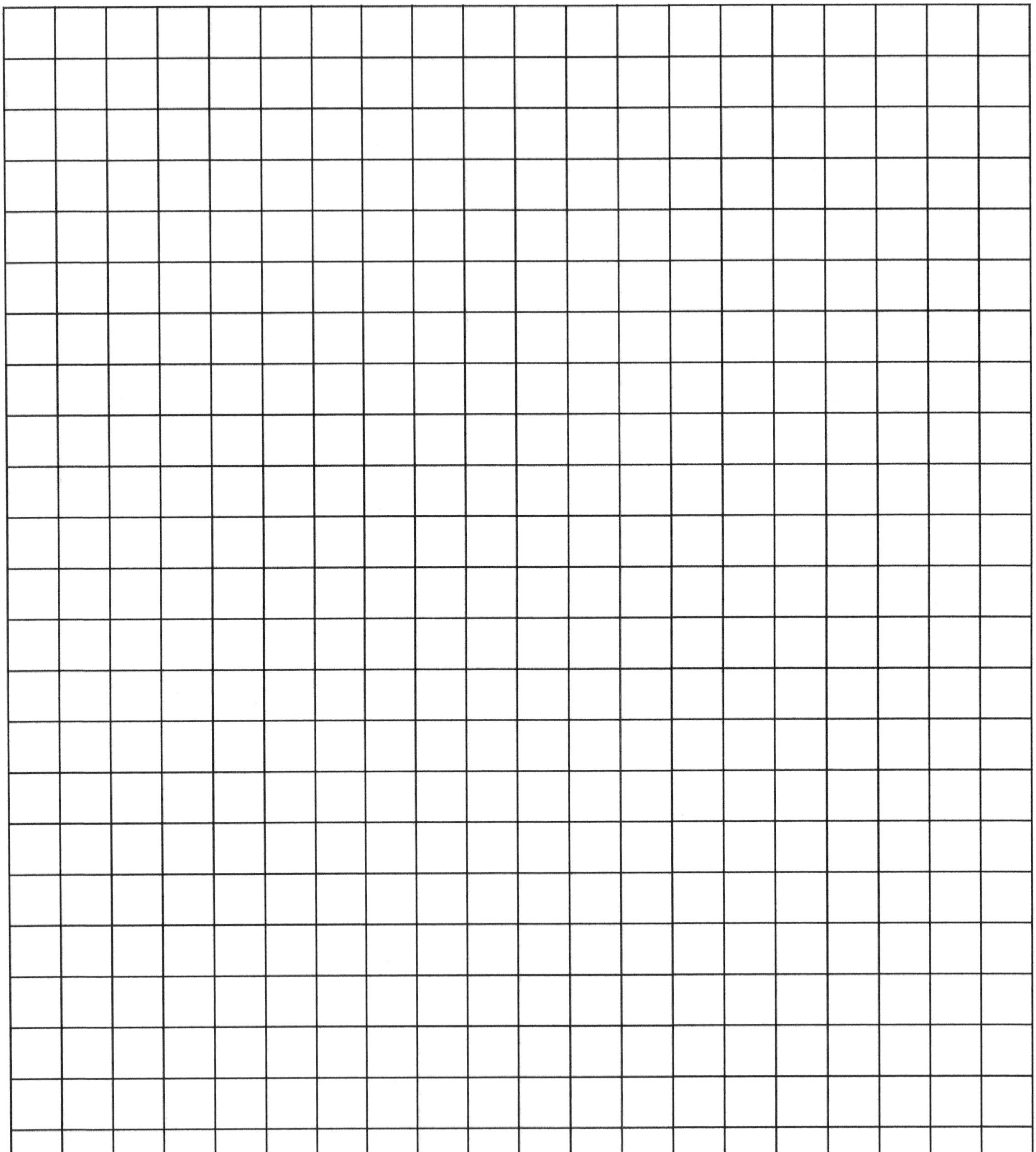

Math Journal

Key Term:

Definition:

Examples and Illustrations:

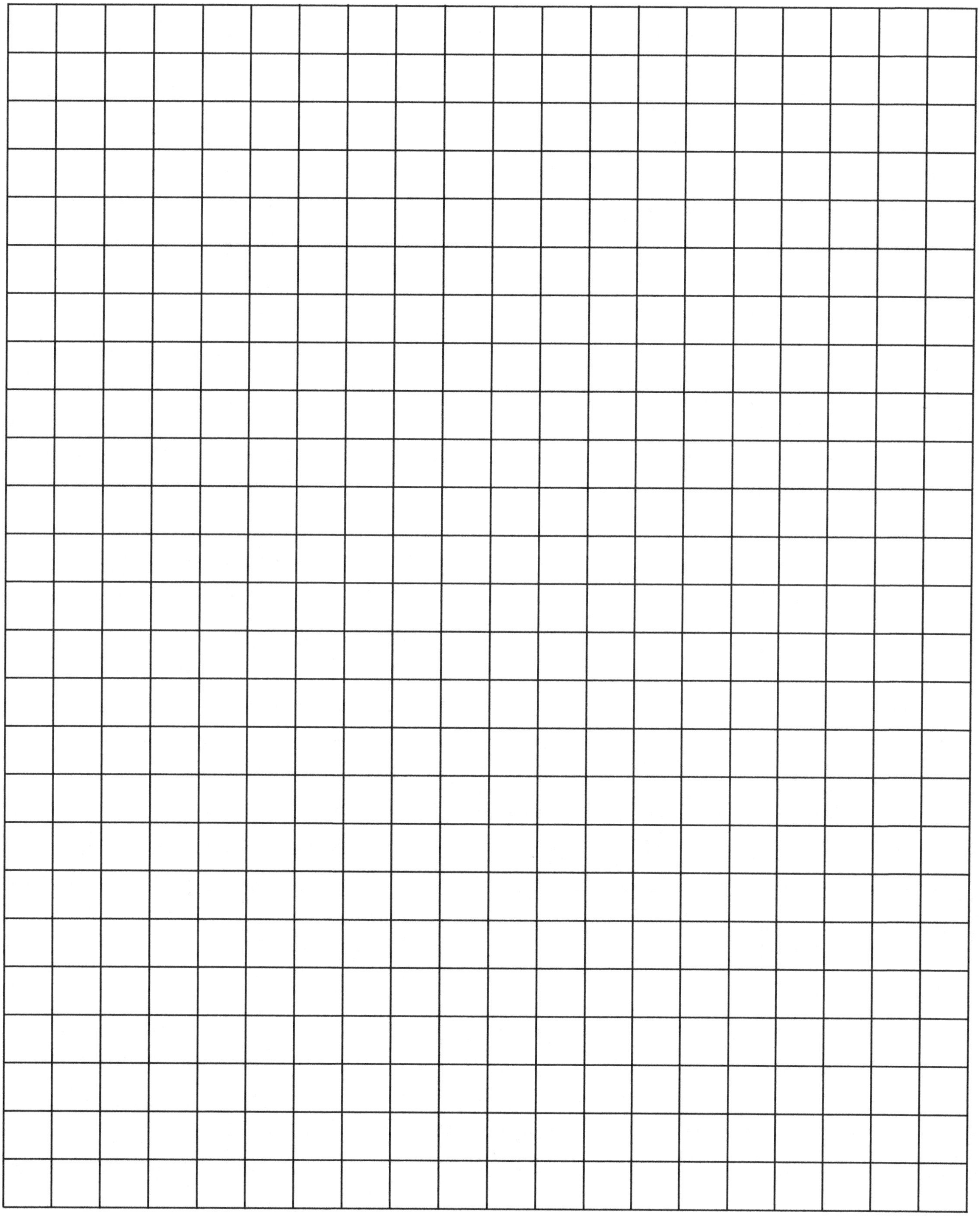

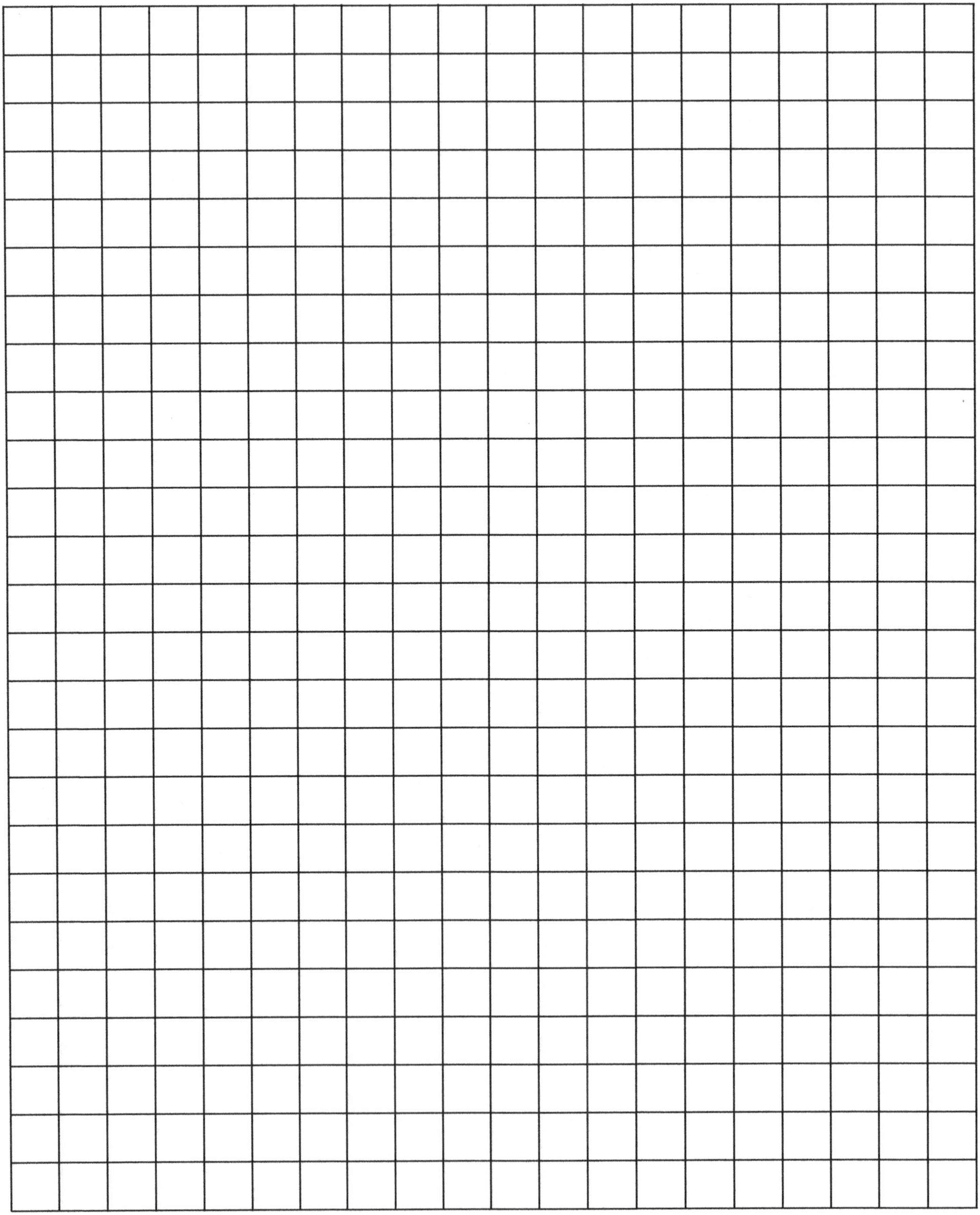

Math Journal

Key Term:

Definition:

Examples and Illustrations:

Math Journal

Key Term:

Definition:

Examples and Illustrations:

Math Journal

Key Term:

Definition:

Examples and Illustrations:

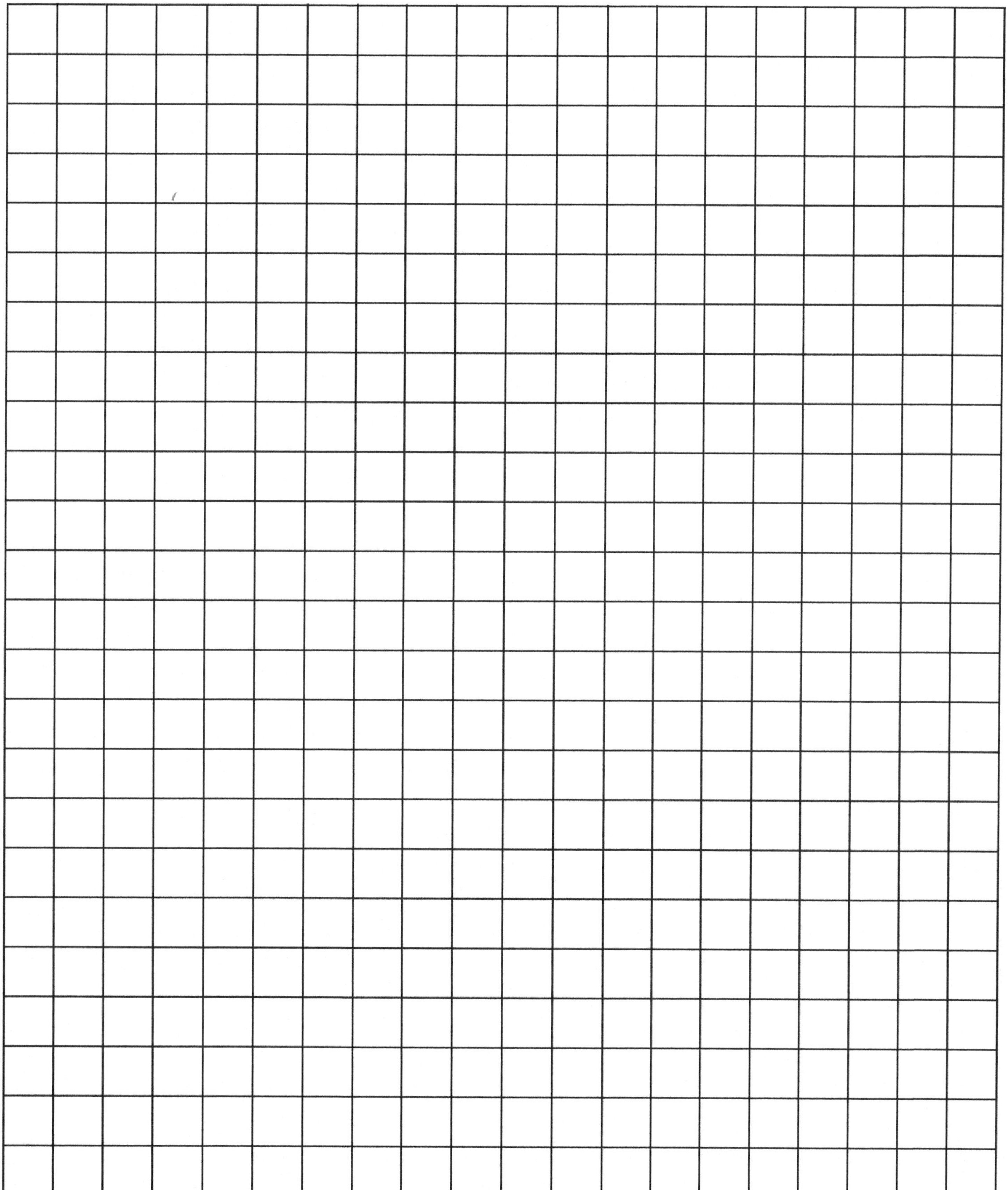

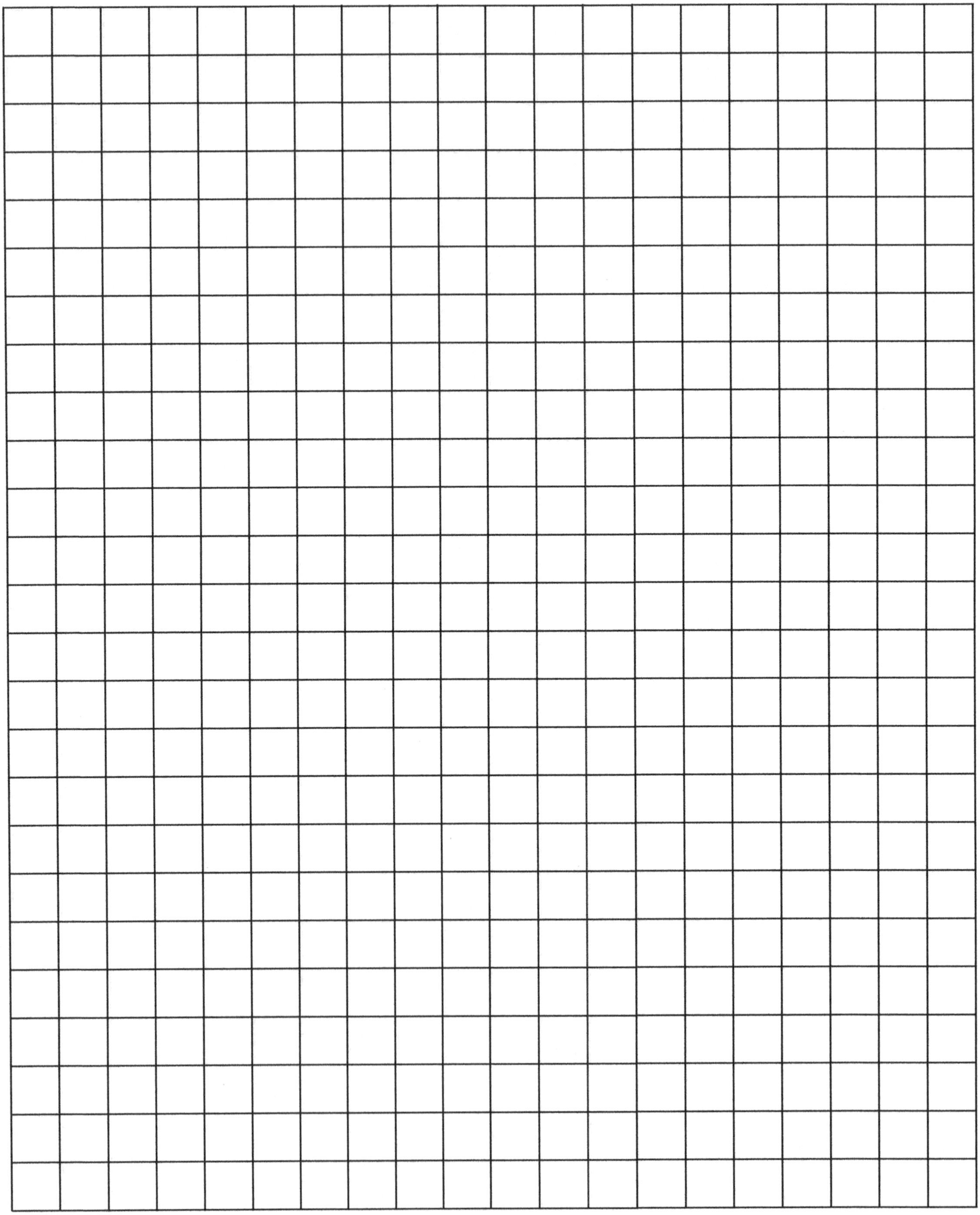

Math Journal

Key Term:

Definition:

Examples and Illustrations:

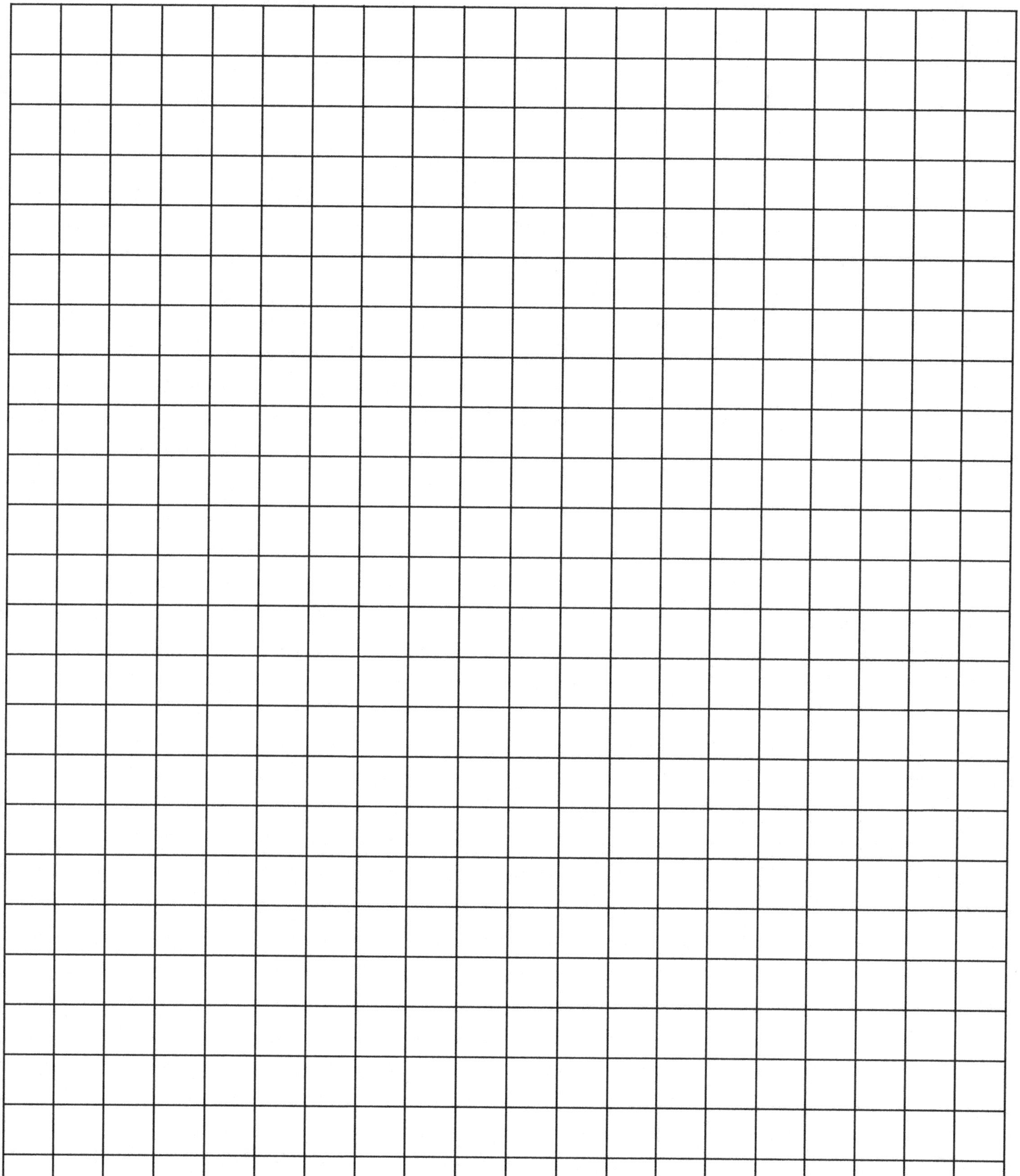

Math Journal

Key Term:

Definition:

Examples and Illustrations:

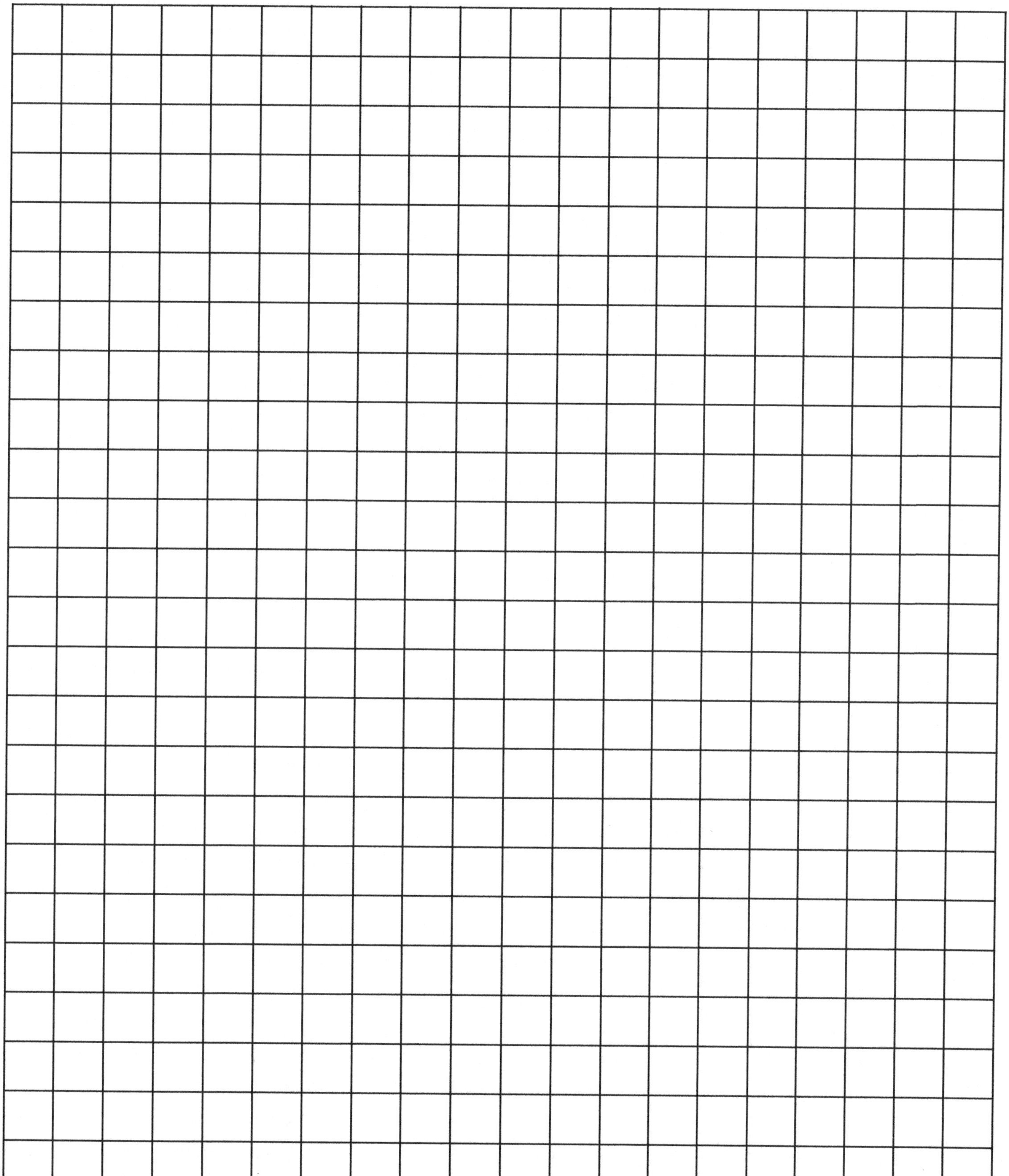

Math Journal

Key Term:

Definition:

Examples and Illustrations:

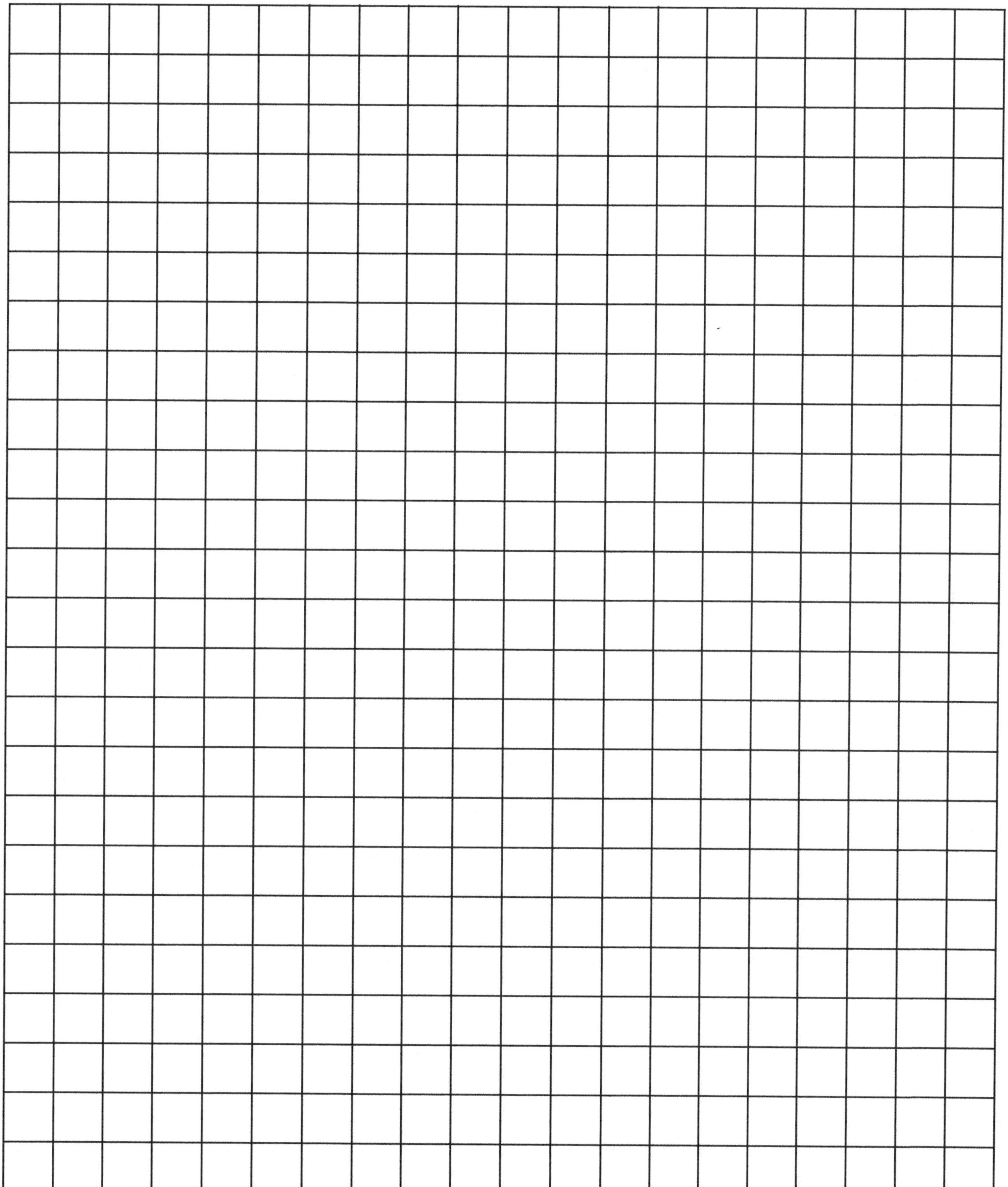

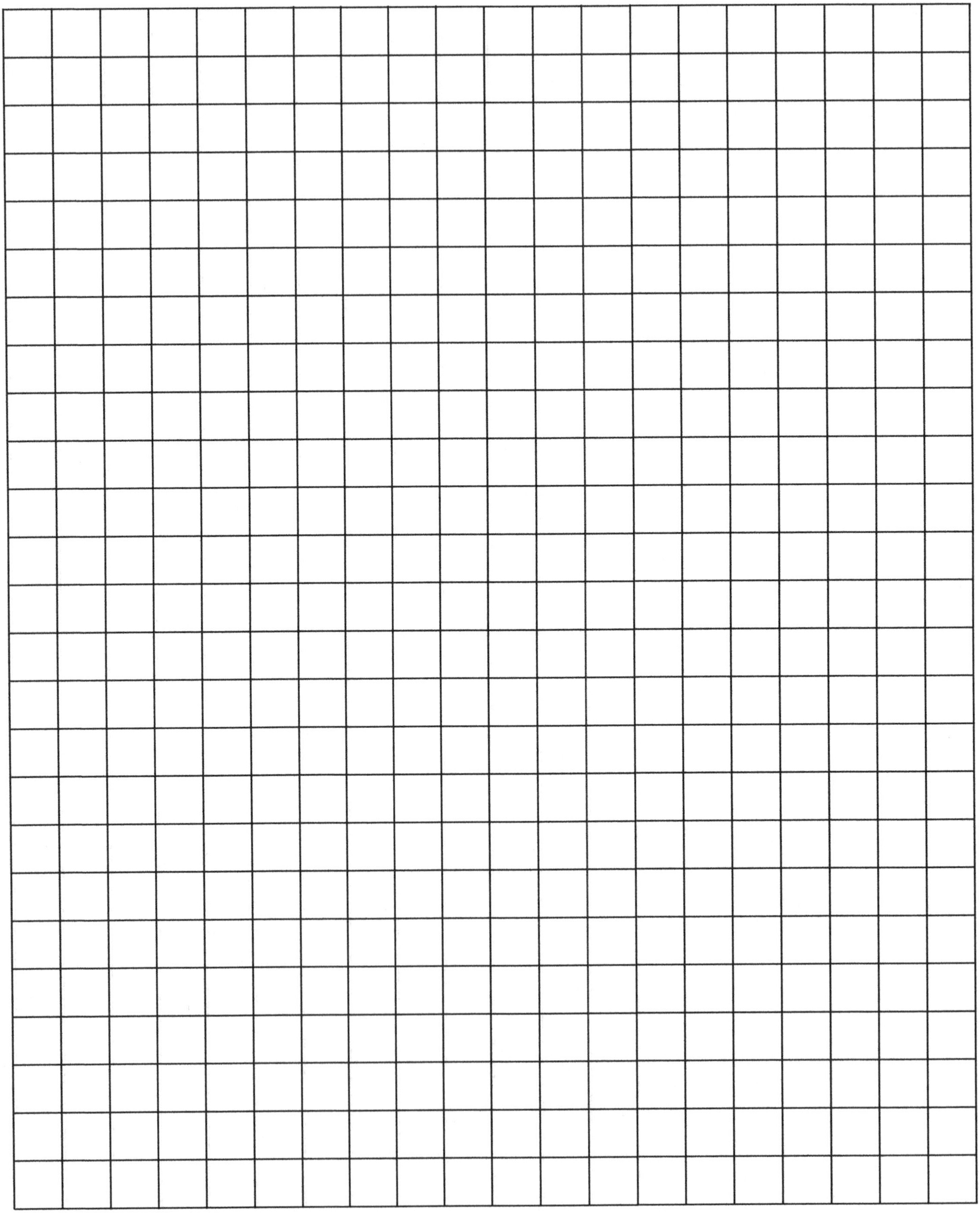

Math Journal

Key Term:

Definition:

Examples and Illustrations:

Math Journal

Key Term:

Definition:

Examples and Illustrations:

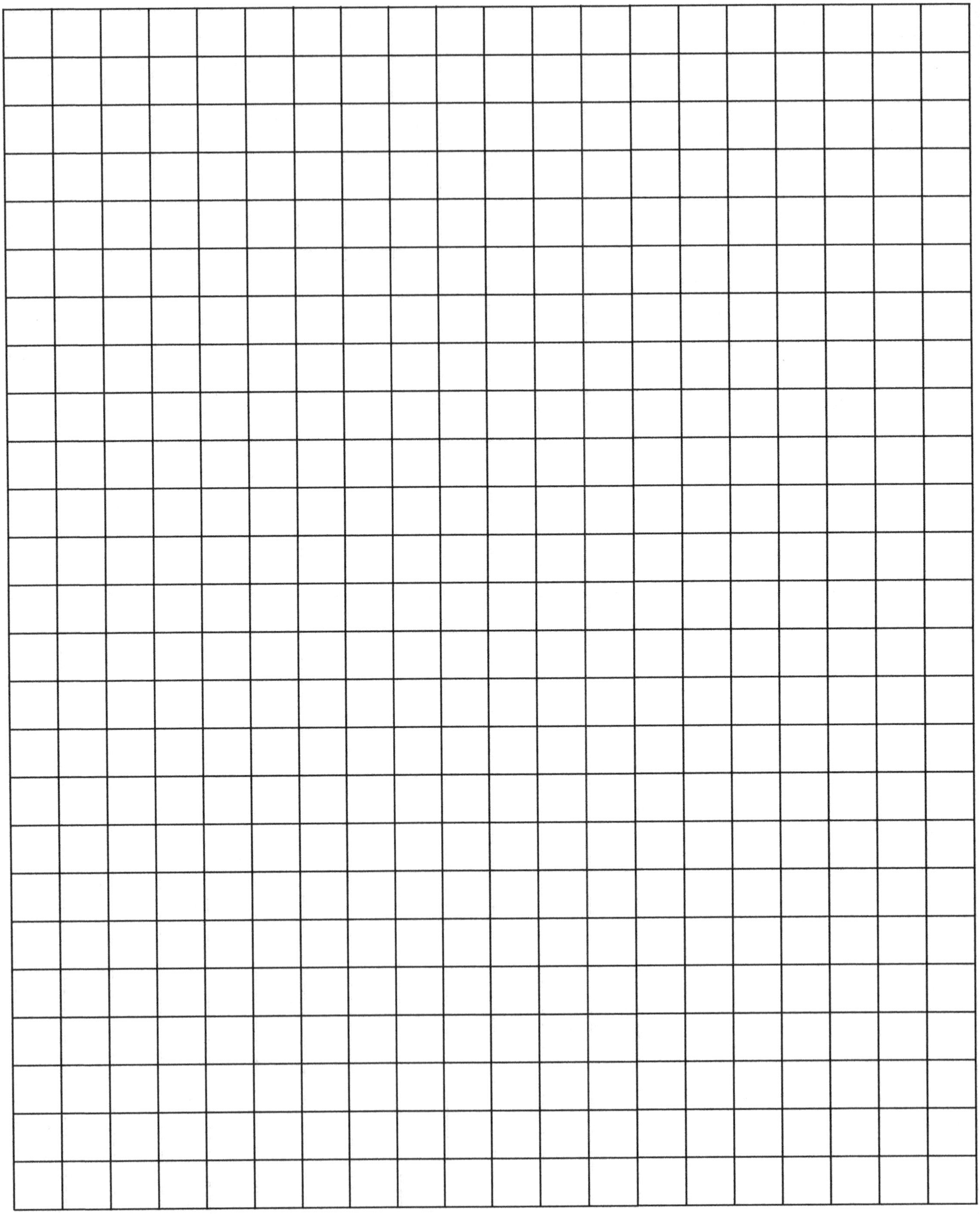

Math Journal

Key Term:

Definition:

Examples and Illustrations:

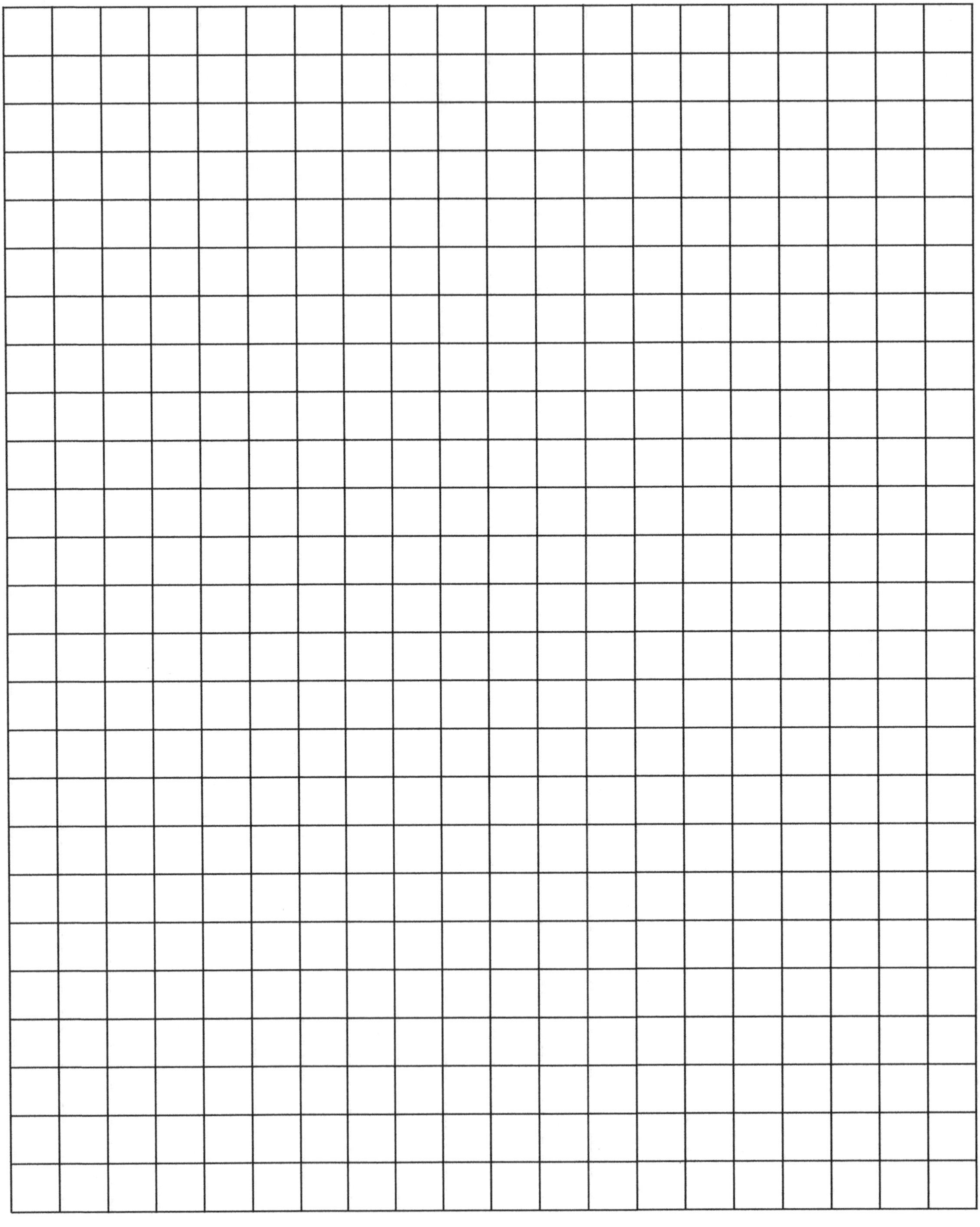

Math Journal

Key Term:

Definition:

Examples and Illustrations:

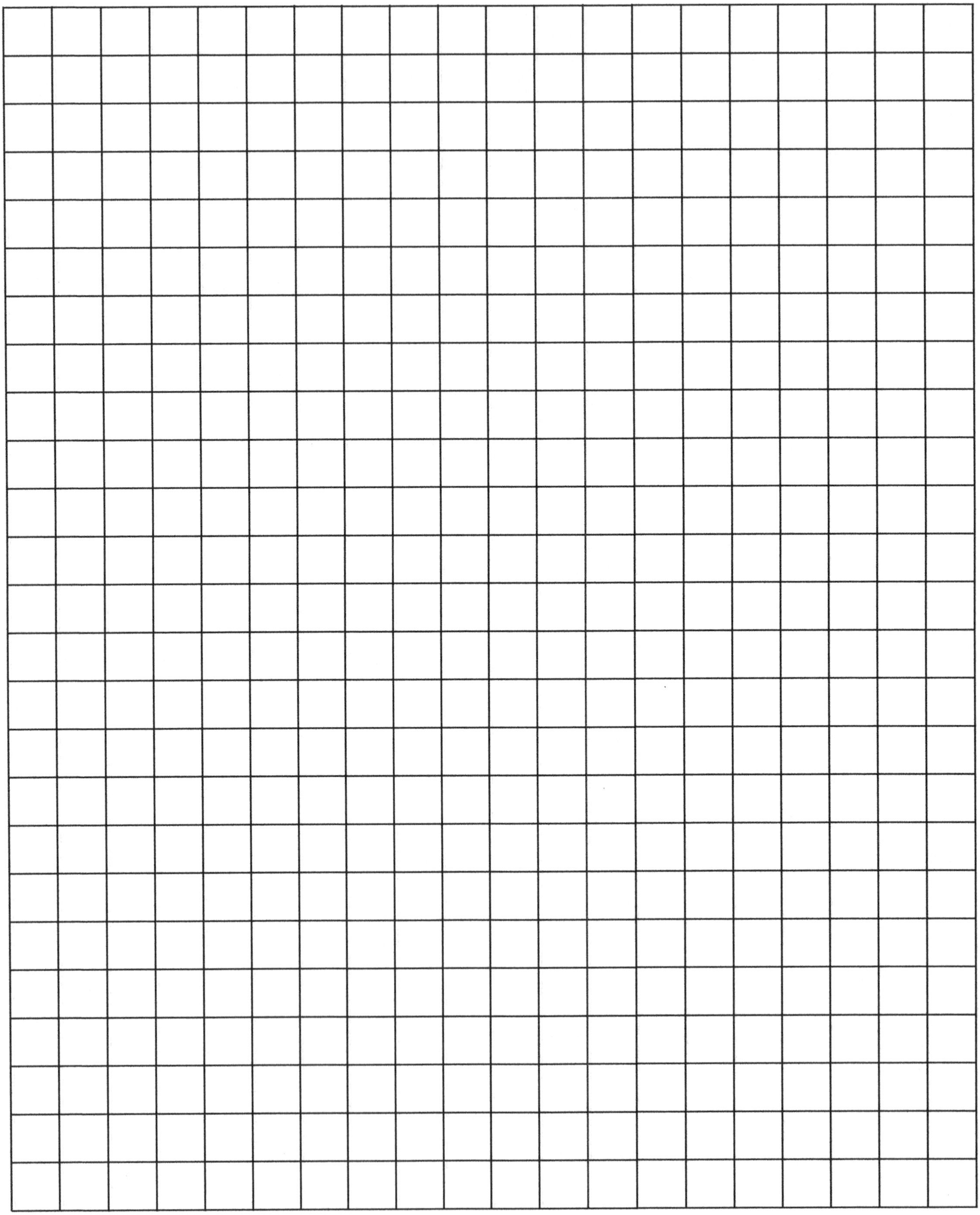

Math Journal

Key Term:

Definition:

Examples and Illustrations:

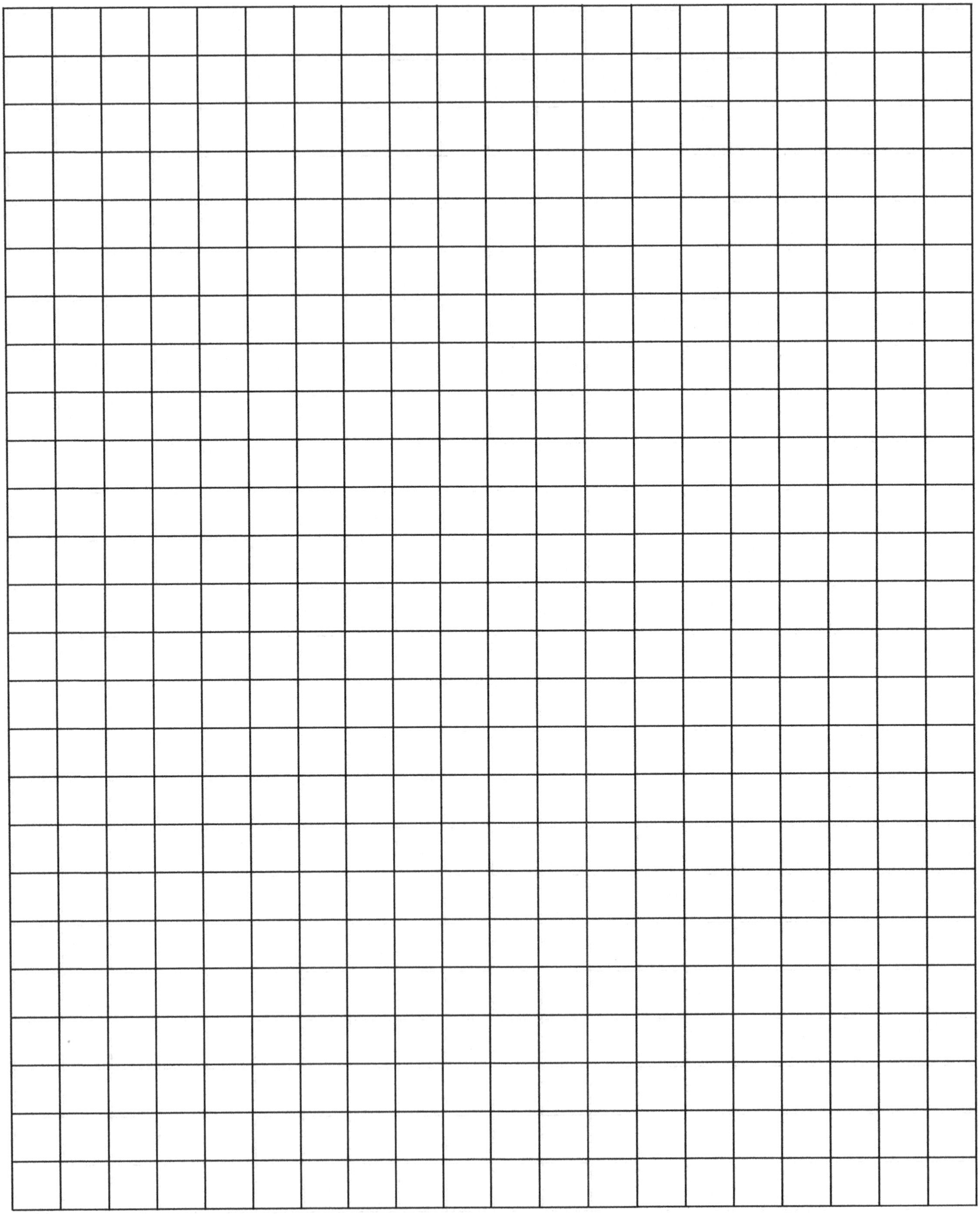

Math Journal

Key Term:

Definition:

Examples and Illustrations:

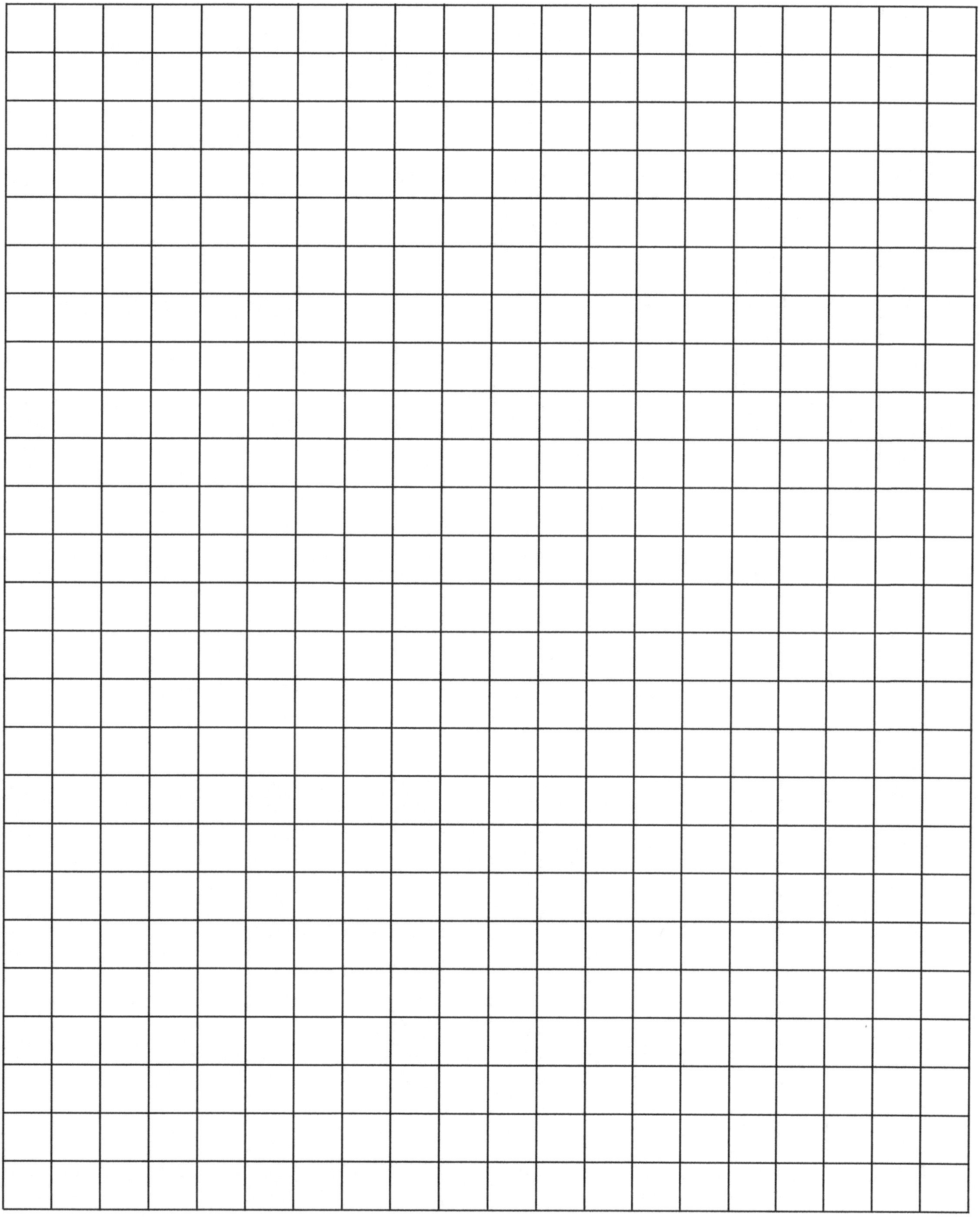

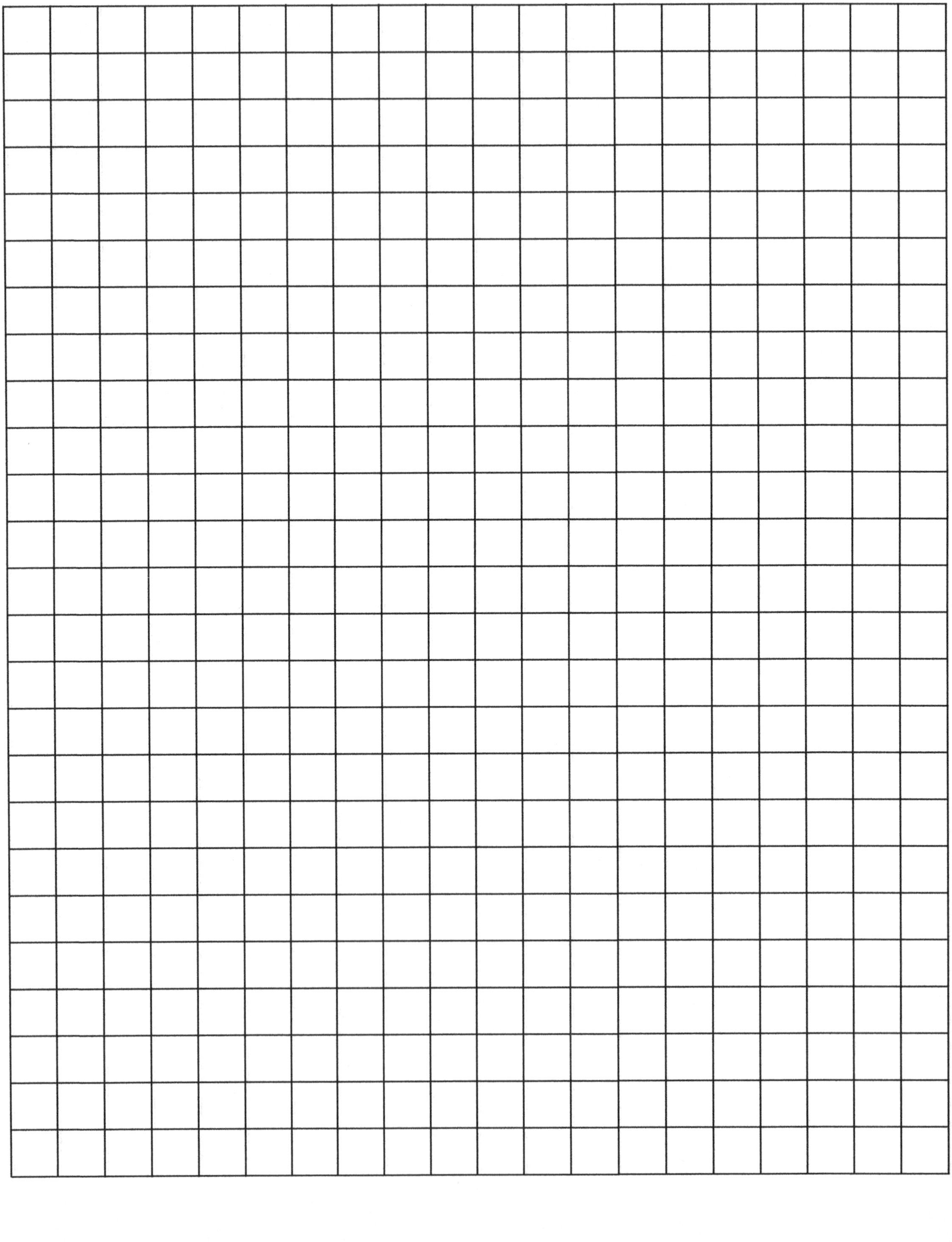

Math Journal

Key Term:

Definition:

Examples and Illustrations:

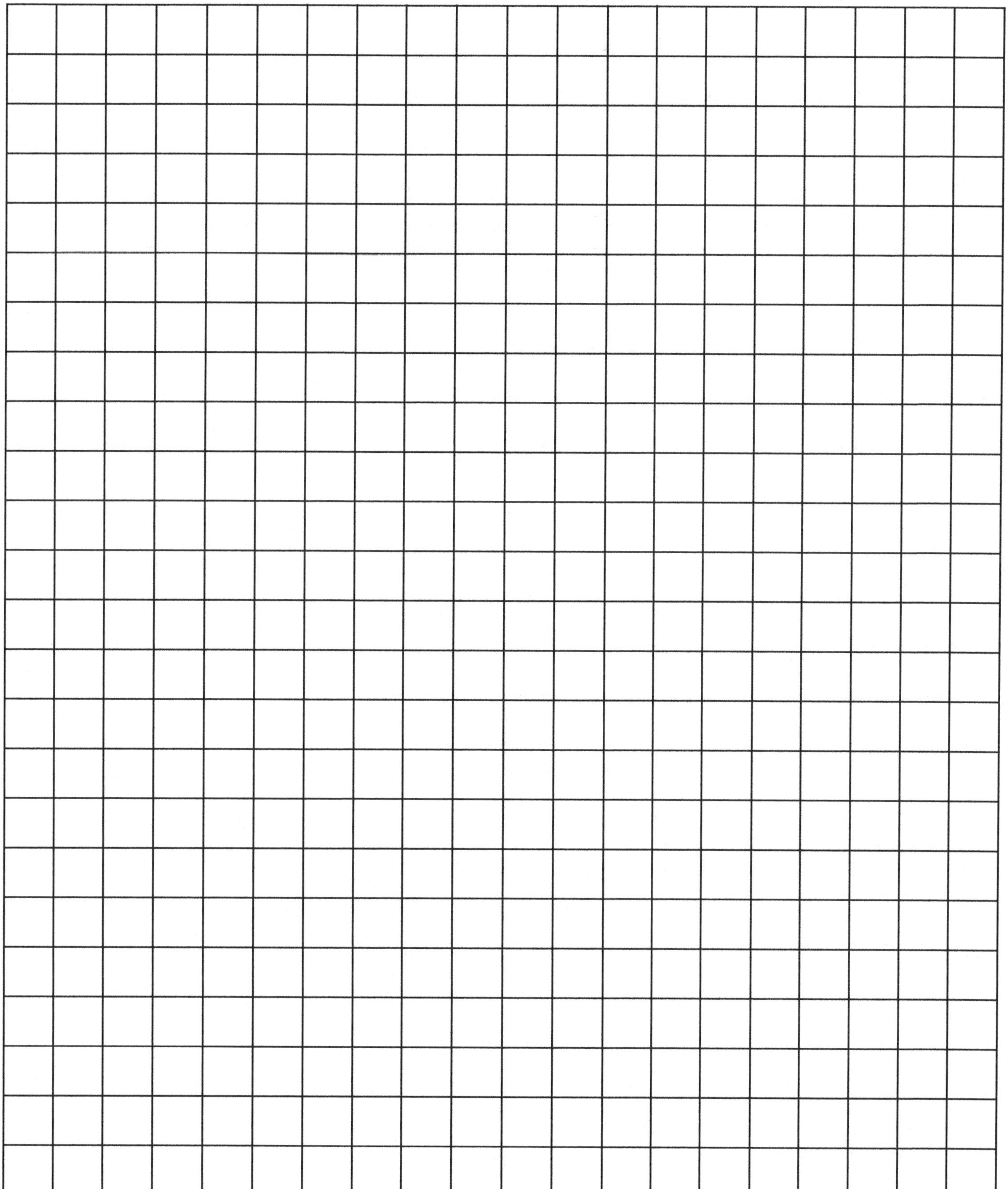

Math Journal

Key Term:

Definition:

Examples and Illustrations:

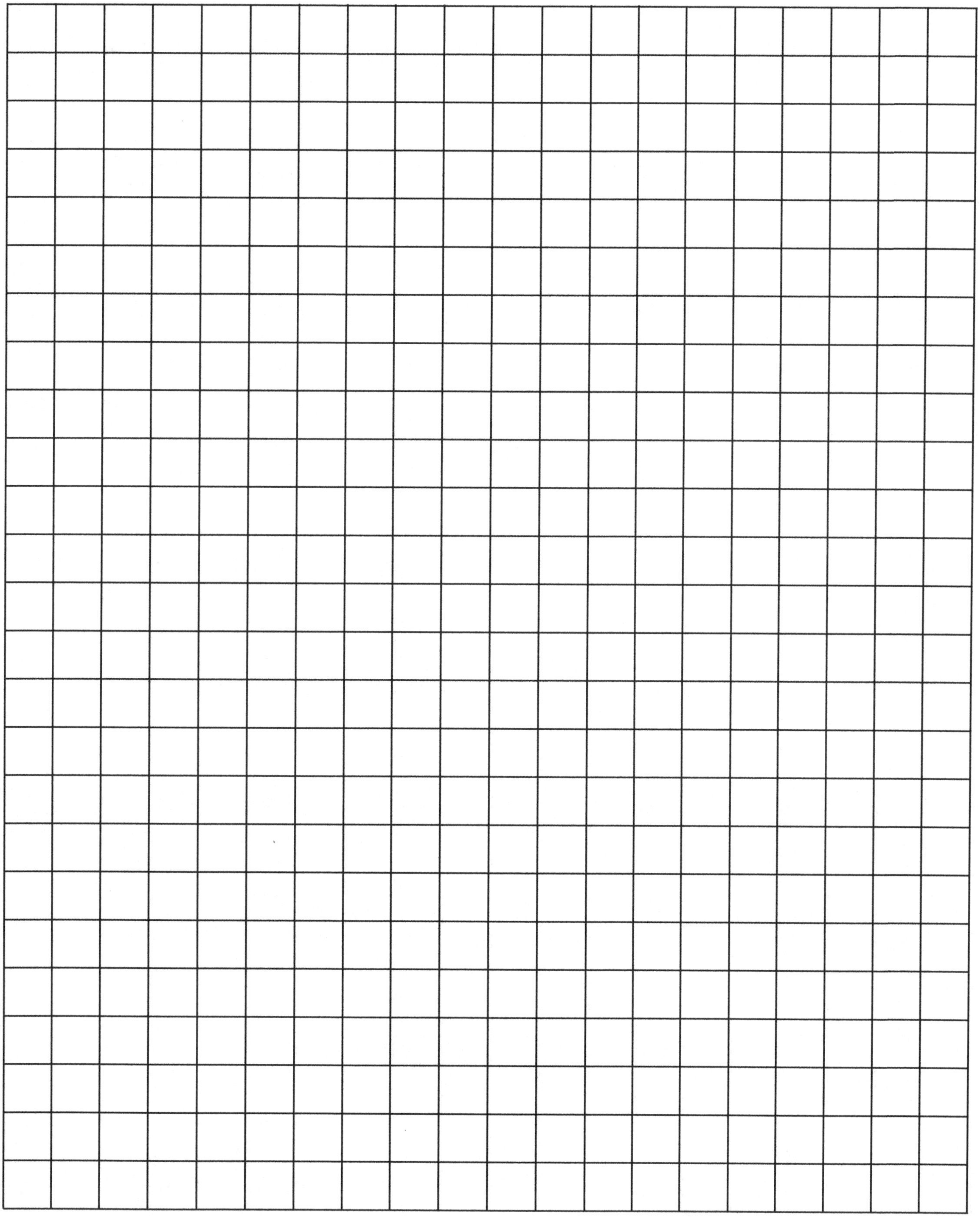

Math Journal

Key Term:

Definition:

Examples and Illustrations:

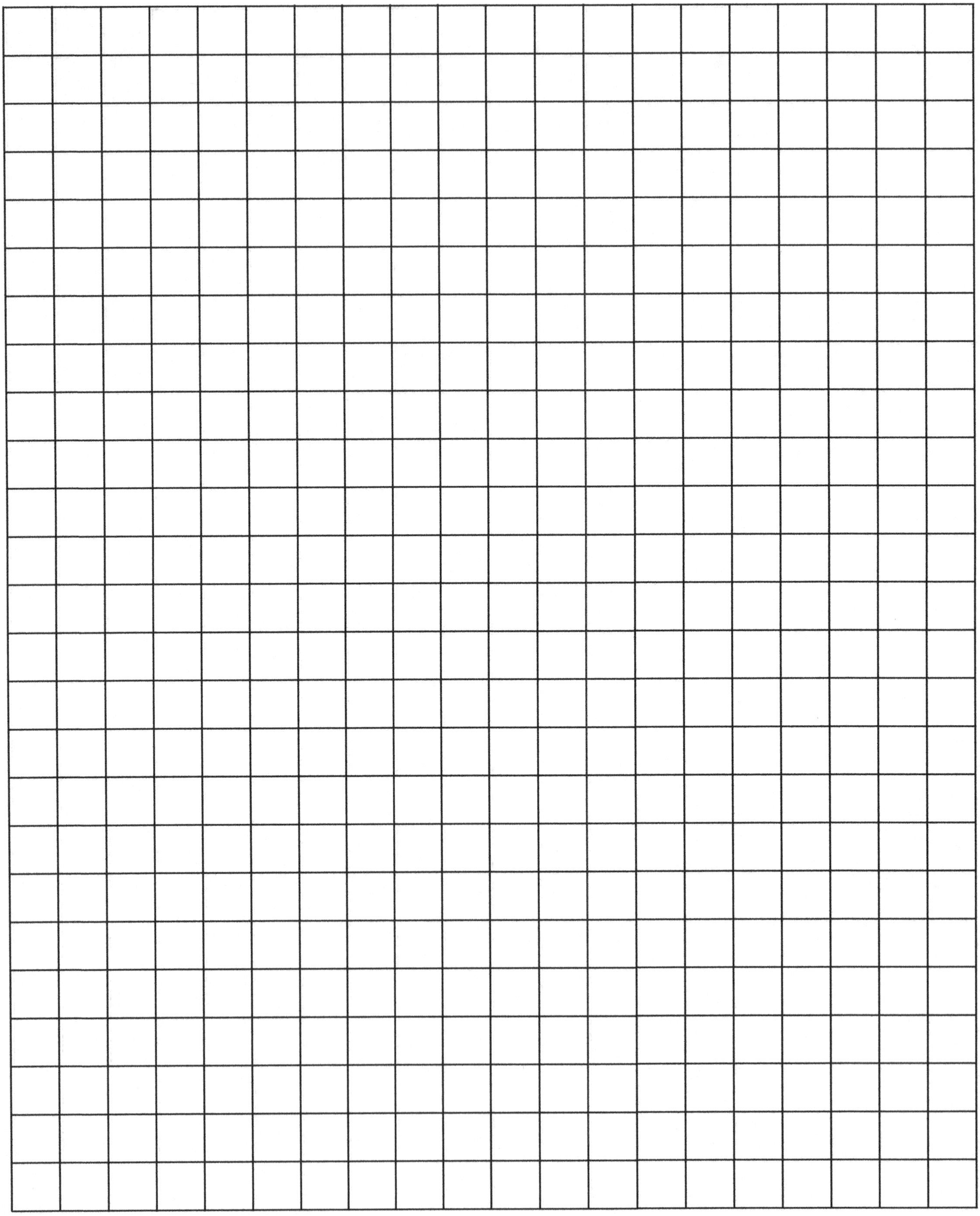

Math Journal

Key Term:

Definition:

Examples and Illustrations:

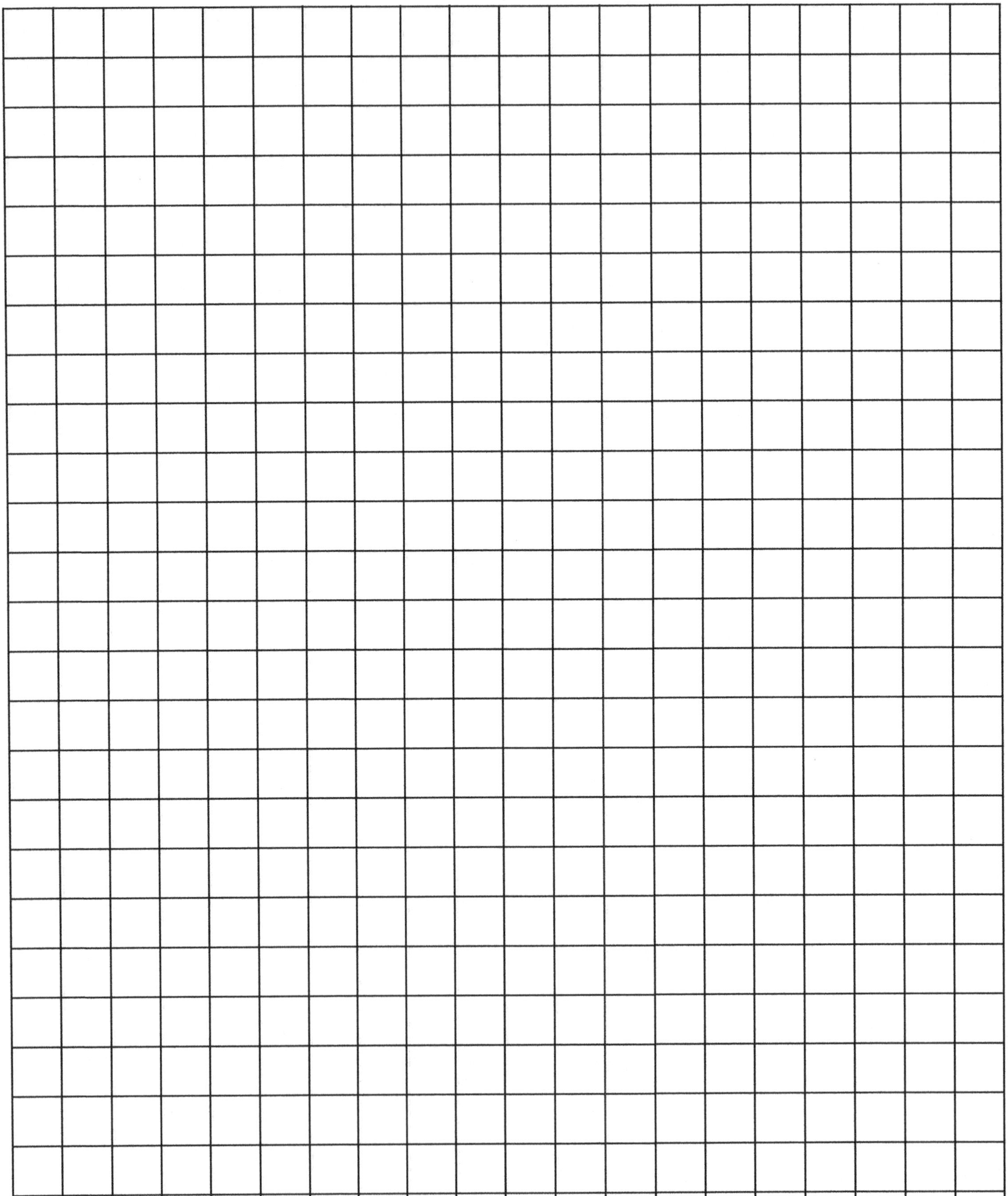

Math Journal

Key Term:

Definition:

Examples and Illustrations:

Multiplication Table

X	1	2	3	4	5	6	7	8	9	10	11	12
1	1	2	3	4	5	6	7	8	9	10	11	12
2	2	4	6	8	10	12	14	16	18	20	22	24
3	3	6	9	12	15	18	21	24	27	30	33	36
4	4	8	12	16	20	24	28	32	36	40	44	48
5	5	10	15	20	25	30	35	40	45	50	55	60
6	6	12	18	24	30	36	42	48	54	60	66	72
7	7	14	21	28	35	42	49	56	63	70	77	84
8	8	16	24	32	40	48	56	64	72	80	88	96
9	9	18	27	36	45	54	63	72	81	90	99	108
10	10	20	30	40	50	60	70	80	90	100	110	120
11	11	22	33	44	55	66	77	88	99	110	121	132
12	12	24	36	48	60	72	84	96	108	120	132	144

Benchmark Fractions

Fraction	Decimal	Percent
$\frac{1}{2}$	.50	50%
$\frac{1}{3}$	.333	33.3%
$\frac{2}{3}$	.666	66.6%
$\frac{1}{4}$	.25	25%
$\frac{3}{4}$	.75	75%
$\frac{1}{5}$	.20	20%
$\frac{2}{5}$	.40	40%
$\frac{3}{5}$	.60	60%
$\frac{4}{5}$	.80	80%
$\frac{1}{6}$	.166	16%
$\frac{1}{8}$	.125	12.5%
$\frac{1}{10}$	.10	10%
$\frac{1}{12}$	.0833	8.3%

Fraction Arithmetic

Add and Subtract

1. Check the Denominator
2. Make them match
3. If you do it to the bottom, do it to the top.
4. Add or subtract the numerators.
5. Keep the denominators the same.
6. Simplify if needed

Multiply

1. Multiply the numerators
2. Multiply the denominators
3. Simplify if needed

Divide

1. Flip the second fraction
2. Multiply the numerators
3. Multiply the denominators
4. Simplify if needed

Rules for the Order of Operations

Please Excuse My Dear Aunt Sally

Parenthesis- Do any operations in parenthesis

Exponents- Do any exponents

Multiplication
Division → Do all multiplication and division in order from left to right

Addition
Subtraction → Do all addition and subtraction in order from left to right

Example:

12 + 4 x (11 – 6)

12 + 4 x 5

12 + 20 = 32

**Remember to only do one step at a time! ☺*

Conversion Chart

Customary Length	
12 in	1 ft
3 ft	1 yd
1760 yd	1 mi

Customary Weight	
16 oz	1 lb
2000 lb	1 ton

Customary Capacity	
8 fl oz	1 c
2 c	1 pt
2 pt	1 qt
4 qt	1 gal

Metric Length	
10 mm	1 cm
100 cm	1 m
1000 m	1 km

Metric Mass	
1000 mg	1 g
1000 g	1 kg

Metric Capacity	
1000 mL	1 L

Customary Length

shorter ⟷ longer

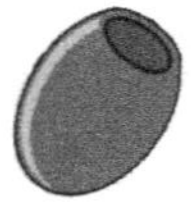

inch	**foot**	**yard**	**mile**
in	**ft**	**yd**	**mi**
olive	foot-long sandwich	baseball bat	4 laps around the track

Customary Weight

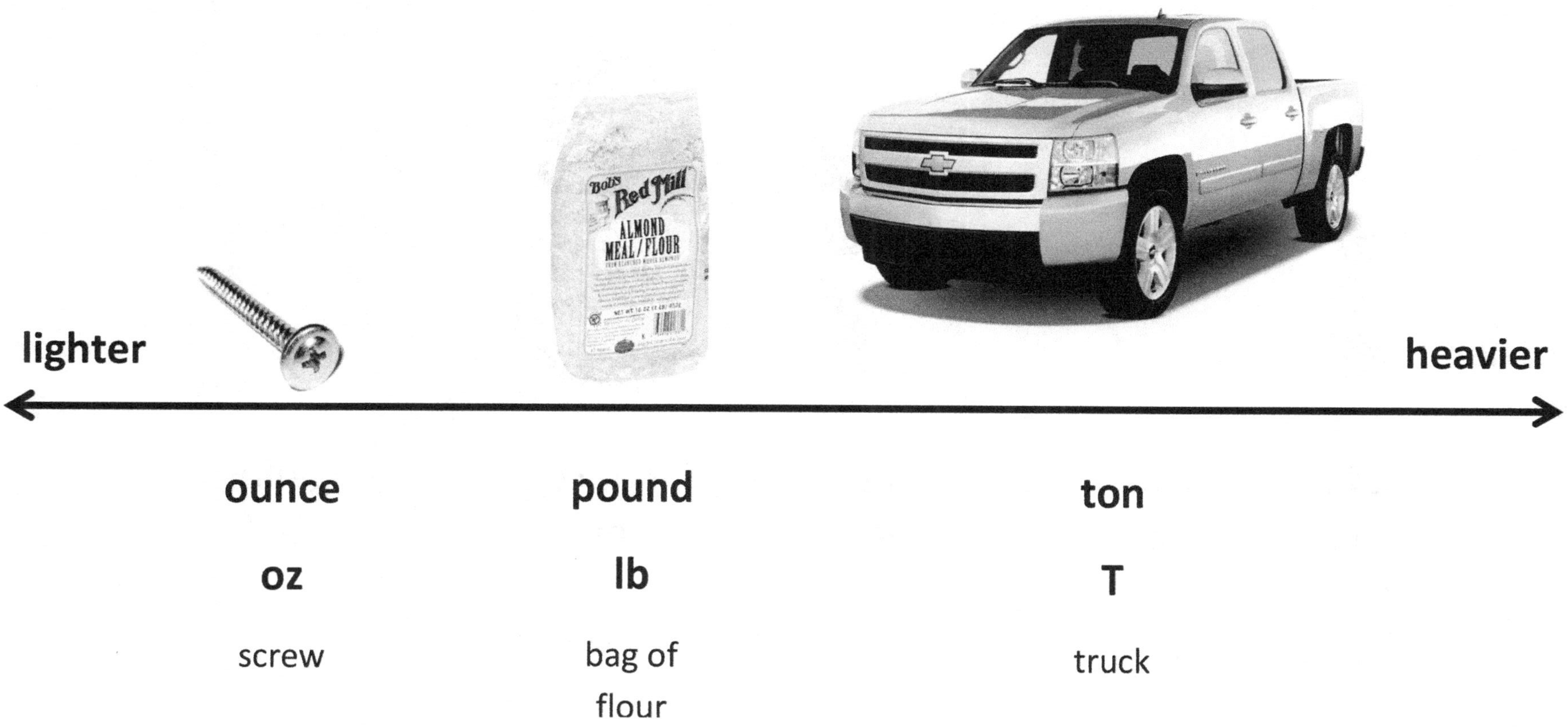

Customary Capacity

smaller ⟵ ⟶ larger

fluid ounce fl oz	**cup c**	**pint pt**	**quart qt**	**gallon gal**
bottle of nail polish	glass of orange juice	small carton of ice-cream	large juice bottle	jug of milk

Metric Length

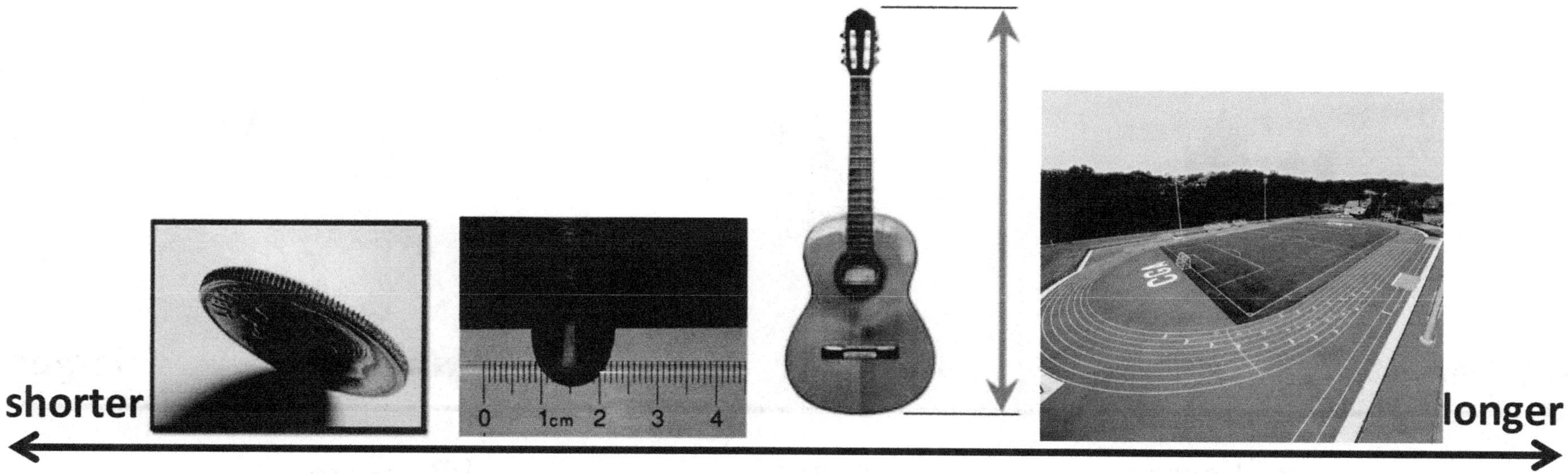

millimeter	**centimeter**	**meter**	**kilometer**
mm	**cm**	**m**	**km**
thickness of a dime	width of your pinky finger	height of a guitar	3 laps around the track

Metric Mass

smaller ⟵⟶ larger

milligram	gram	kilogram
mg	g	kg
a leaf	a thumbtack	a genuine gold coin

Metric Capacity

smaller ⟷ larger

milliliter
mL

a dropper

liter
L

a large soda bottle

Integer Rules

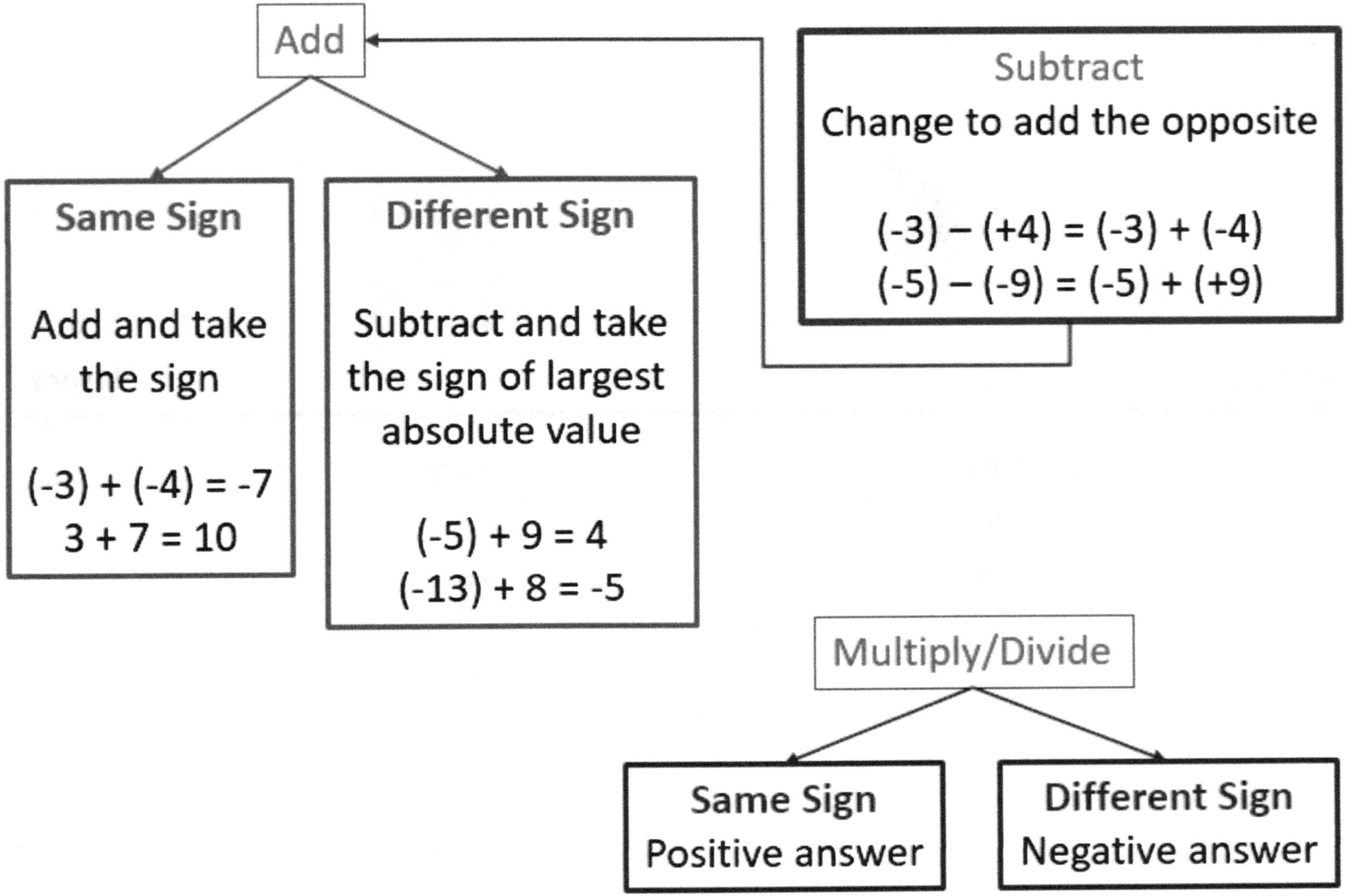

Area, Volume, and Trigonometry Formulas

Area of a Rectangle or a Parallelogram

$A = bh$

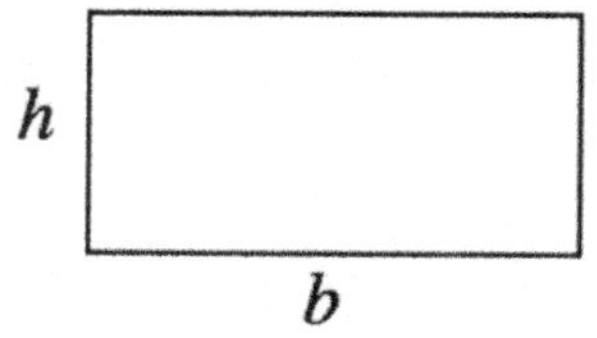

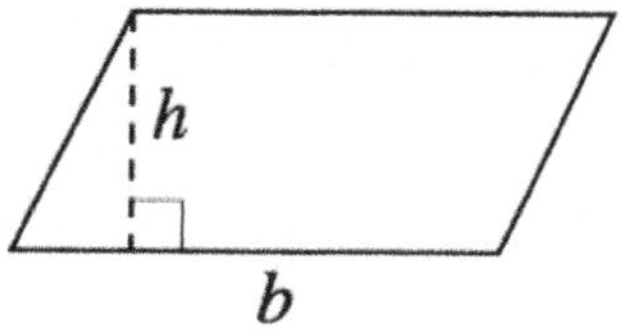

Area of a Triangle

$A = \frac{1}{2} bh$

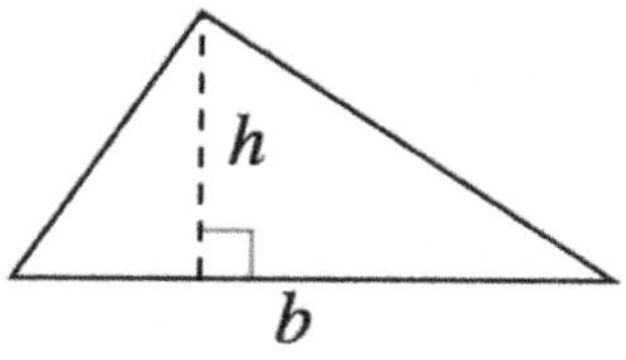

Area of a Trapezoid

$A = \frac{1}{2} h(b_1 + b_2)$

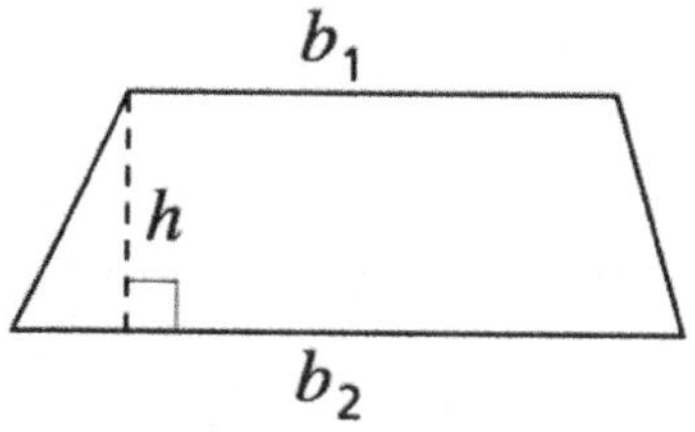

Area of a Circle and Circumference of a Circle

$A = \pi r^2$

$C = 2\pi r$

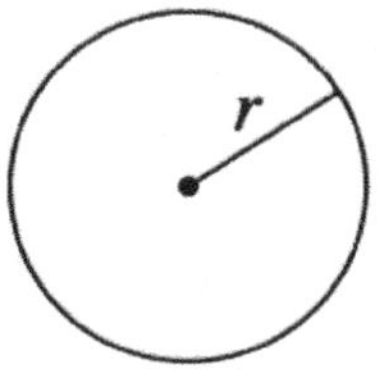

Length of an Arc of a Circle and Area of a Sector of a Circle

$s = \frac{m}{360} C$

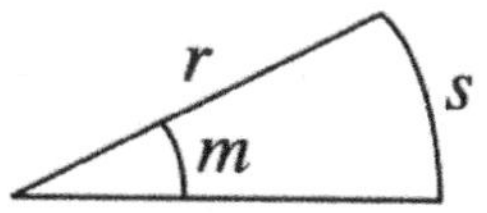

$A_{\text{sector}} = \frac{m}{360} \pi r^2$ where m is the angle measure in degrees

Volume of a Prism or a Cylinder

$V = Bh$

where B is the area of the base

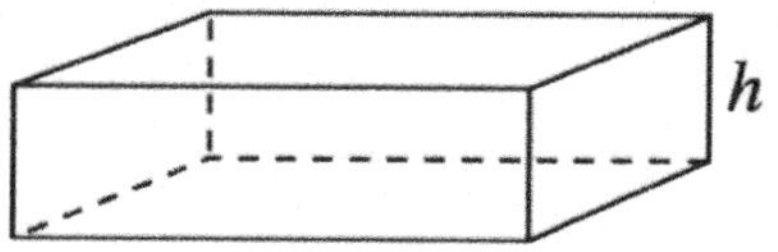

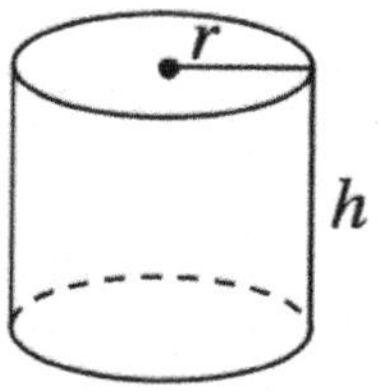

Pythagorean Therorem

$a^2 + b^2 = c^2$

Distance Formula

$d = \sqrt{(x_2 - x_1)^2 + (y_2 - y_1)^2}$

Quadratic Formula

$x = \frac{-b \pm \sqrt{b^2 - 4ac}}{2a}$

Trigonometric Relations

$\sin\theta = \frac{\text{opposite}}{\text{hypotenuse}}$ $\cos\theta = \frac{\text{adjacent}}{\text{hypotenuse}}$ $\tan\theta = \frac{\text{opposite}}{\text{adjacent}}$

Made in the USA
Monee, IL
08 November 2026

67377508R00063